AHA! ACADEMY

SHINING BRIGHT!

The Physics of Light

Written by Angela Lim

WORLD BOOK

www.worldbook.com

Co-published by agreement between Shi Tu Hui and World Book, Inc.

Shi Tu Hui
Room 1807, Block 1,
#3 West Dawang Road
Chaoyang District, Beijing 100025
P.R. China

World Book, Inc.
180 North LaSalle Street
Suite 900
Chicago, Illinois 60601
USA

Library of Congress Control Number: 2024947054

Aha! Academy: Physics
ISBN: 978-0-7166-7144-2 (set, hard cover)

Get Moving! The Physics of Motion
ISBN: 978-0-7166-7147-3 (hard cover)
ISBN: 978-0-7166-7167-1 (e-book)
ISBN: 978-0-7166-7157-2 (soft cover)

Printed in India by Replika Press PVT LTD, Haryana, India
1st printing January 2025

Staff

Editorial

Vice President
Tom Evans

Editorial Project Coordinator
Kaile Kilner

Senior Curriculum Designer
Caroline Davidson

Proofreader
Nathalie Strassheim

Graphics and Design

Senior Visual
Communications Designer
Melanie Bender

Designer
Shannon Hagman

Digital Asset Specialist
Rosalia Bledsoe

Written by Angela Lim
Advised by Julie Gunderson

Developed with World Book by
Red Line Editorial

Acknowledgments

The publishers gratefully acknowledge the following sources for photography. All illustrations were prepared by WORLD BOOK unless otherwise noted.

Cover: improvise/Shutterstock; Kartinkin77/Shutterstock; khlungcenter/Shutterstock; Pix Cell/Shutterstock; Tatevosian Yana, Shutterstock

Addictive Stock Creatives/Alamy Images 15; Simon Belcher, Alamy Images 4; Robert Harding, Alamy Images 10; Mint Images Limited/Alamy 4; NASA, ESA, and D. Coe, J. Anderson, and R. van der Marel (STScI) 43; Public Domain (Shinobu Ishihara) 29; Shutterstock 3, 4, 5, 6, 7, 8, 9, 10, 11, 12, 13, 14, 15, 16, 17, 18, 19, 20, 21, 22, 23, 24, 25, 26, 27, 28, 29, 30, 31, 32, 33, 34, 35, 36, 37, 38, 39, 40, 41, 42, 43, 44, 45, 46, 47, 48; Wellcome Images (licensed under CC BY 4.0) 40

There is a glossary of terms on page 48. Terms defined in the glossary are in type that looks like *this* on their first appearance on any spread (two facing pages).

Contents

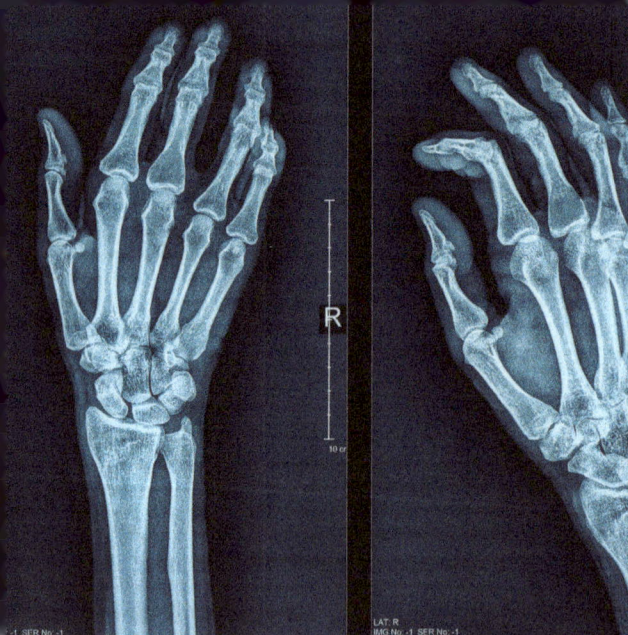

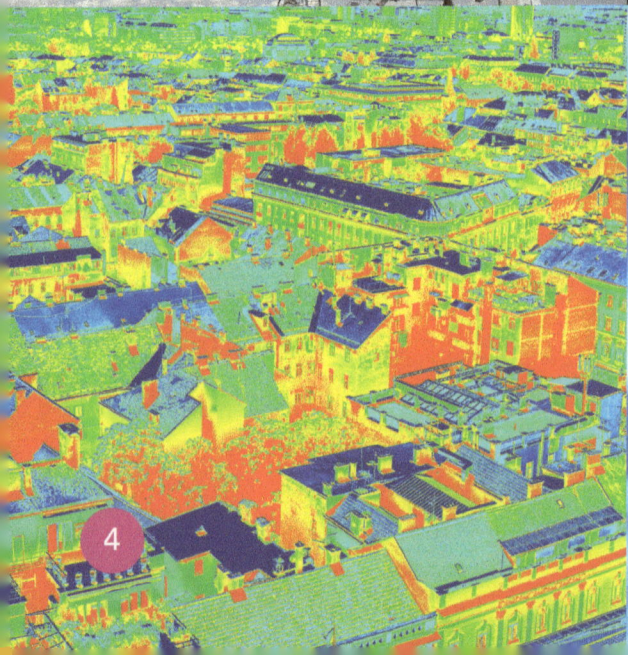
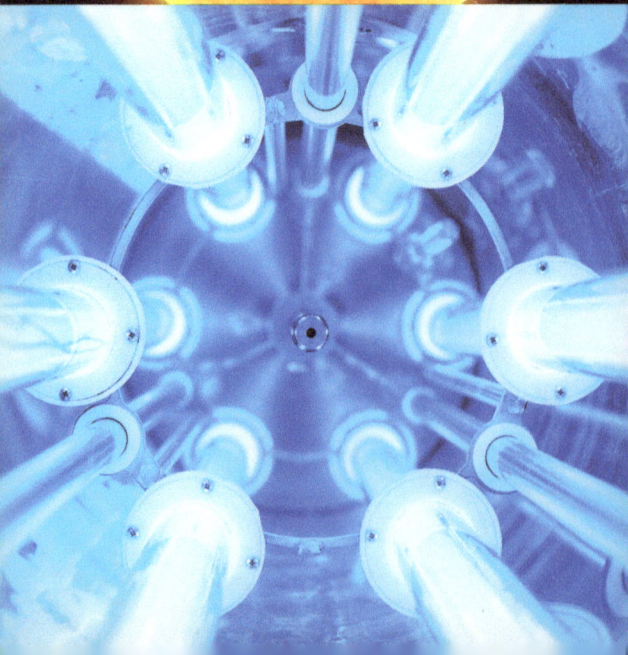

4

Introduction

Can you imagine life without light? In short, it wouldn't be possible! The sun is a major source of the light we see. But visible light is just one type of light that the sun emits. It also gives off invisible forms of light. For example, your skin gets sunburned because of invisible light called ultraviolet (UV) rays.

Scientists study the physics of light—that is, what it's made of and how it behaves. And do you know one thing they've discovered? Light is the fastest thing in the universe!

Researching light helps us understand so much about nature, outer space, and even how our bodies work. It also leads to all kinds of innovations. Read on to learn more about how light impacts our daily lives!

You can listen to the radio thanks to radio waves, another form of invisible light!

1

LET THERE BE LIGHT!

Light is made of particles of energy called *photons*. Photons interact with materials and transfer energy.

The blazing sun. A brilliant flash of lightning. A crackling bonfire. A flickering light bulb. All of these are sources of light! But what *is* light?

DID YOU KNOW?

Staring at the sun and other bright sources of light damages the eyes. It can even cause blindness over time. Even when wearing sunglasses, you shouldn't look directly at the sun.

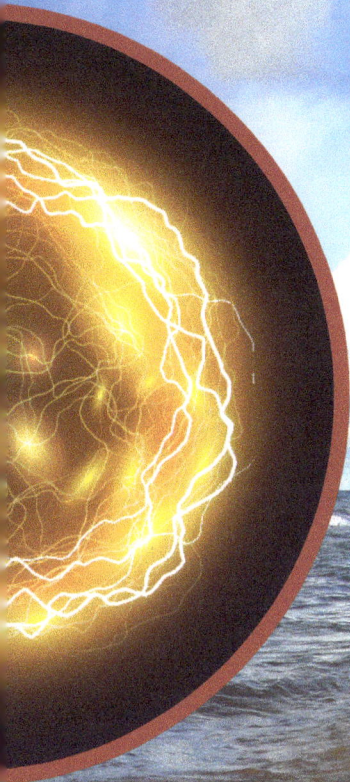

Some light sources are natural. The sun and other stars, for example, are natural sources of light. Sunlight must travel a long way to reach Earth—about 93 million miles (150 million kilometers)!

Animals can also be natural sources of light! For example, the lower part of a firefly's body gives off a yellow glow. These insects use light to find mates.

Other light sources are artificial. This means people made them. Candles and light bulbs are examples of artificial light sources.

Imagine a wave rippling across a lake. The wave slows down and scatters once it hits an object, such as a rock or log. Light behaves the same way. It reacts to objects in its path.

Getting **particular**

Imagine you are at a concert with a bunch of spotlights. These spotlights may cross paths. The beams of light pass through each other. The light behaves this way because photons do not have *mass*. If photons had mass, beams of light would bounce off each other and scatter!

Beeeeep!

TECH TIME

Solar panels work by capturing the energy of photons. Sunlight shines on the panel, and energy from the photons ejects *electrons* from the panel. This creates an electrical current.

Photons **are the particles that make up light.**

These itty-bitty bundles of energy zip and zoom all around us.

Though photons are tiny and massless, they can play a big role in technology! For example, some smoke detectors use light to sense smoke. A light inside the smoke detector shines in a straight beam. When smoke particles enter the device, they scatter the beam of light in all directions. When photons in the scattered light hit a sensor, it causes an alarm to sound.

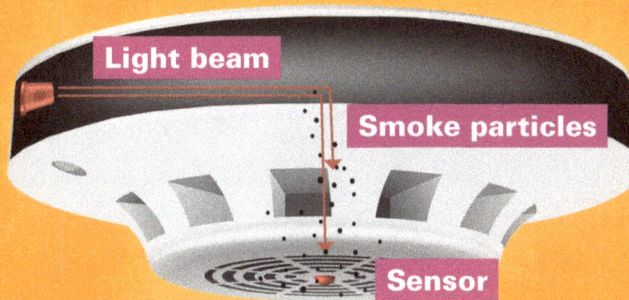

Light beam

Smoke particles

Sensor

DO NOT PAINT

Zooming right along

Lightning flashes. A few moments later, thunder crashes. Why is there a delay before we hear thunder? Because light travels about a million times faster than sound! This is why there is a longer pause between seeing lightning and hearing thunder when the storm is farther away.

DID YOU KNOW?

If a star is 1,000 light-years from Earth, that means the light we see coming from it is 1,000 years old. Looking at the stars is almost like looking back in time!

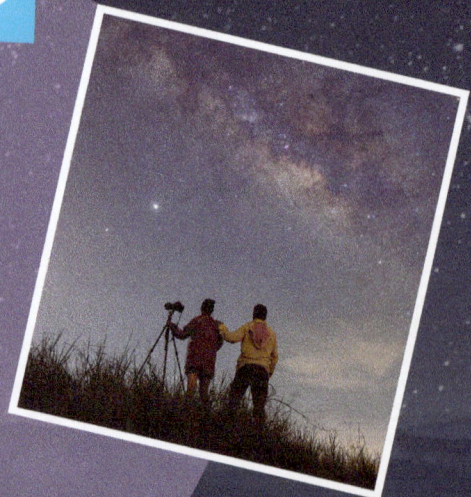

Light travels at nearly 984 million feet (300 million meters) per second. If you could move at the speed of light, it would take you only 0.13 seconds to circle Earth!

Light can be used to measure distance.

A light-year is the distance light travels in a single year. This is about 6 trillion miles (9.6 trillion km). Meanwhile, a light-second—the distance light travels in a single second—is about 186,000 miles (300,000 km)!

1.3 light-seconds

Moon

The moon *reflects* light from the sun.

8 light-minutes

Sun

The sun is the closest star to Earth.

4.24 light-years

Proxima Centauri

Proxima Centauri is the second closest star to Earth.

26,000 light-years

Center of Milky Way

Earth is part of a cluster of stars and planets called the Milky Way galaxy. In some places, we can see the Milky Way from Earth!

The dark side

Certain factors affect how a shadow looks. The shape of the object matters. A round object has a round shadow. A person has a shadow that looks like them.

Let's go, shadow!

The brightness of a light affects the way a shadow looks. An object that blocks a very bright light will cast a dark shadow. A shadow from a dim light is not as dark.

The distance between the light source and the object also changes what the shadow looks like. An object that is close to the light source has a shadow with defined edges. An object that is far away from the light source casts a blurry shadow.

12

Imagine you're shining a flashlight into a dark room. The light travels outward from its source. That is, until something gets in its way! When an object blocks light, a shadow is formed.

The material an object is made of also affects its shadow. Objects can be *opaque*, *translucent*, or *transparent*. These terms describe how much light can pass through the material.

No light can pass through opaque objects. Because of this, opaque objects have dark shadows. People, books, and trees are examples of opaque objects.

Light partially passes through translucent objects. As a result, these objects cast a dim shadow with a blurry outline. Waxed paper and stained glass are examples of translucent materials.

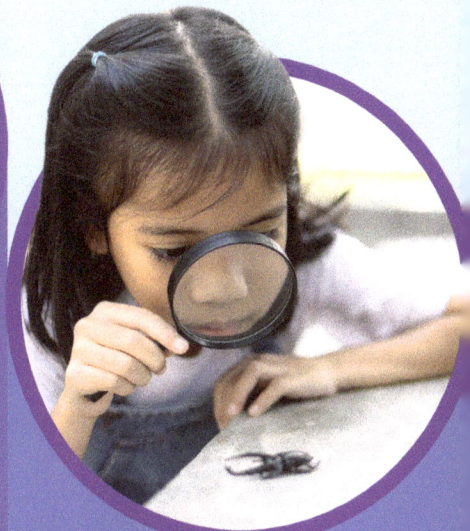

Light can fully pass through transparent objects, so these objects do not have shadows. Water, plastic, and glass are transparent materials.

Lighting the **way**

Absorption

Most of the light that strikes an *opaque* object is *absorbed*. Light energy is transformed into heat when it is absorbed.

Reflection

Reflection occurs when light bounces off an object. For example, when light strikes a mirror, the light reflects off the mirror and travels in a new direction.

Refraction

Light slows down as it travels from air into water. This change in speed causes a slight change in direction called *refraction*. Refraction is the reason a straw in a glass of water looks bent!

14

When light encounters *matter*, it interacts with the substance. Let's explore the ways in which light interacts with the world around us!

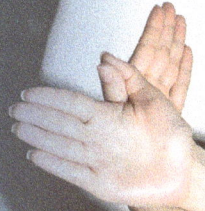

Look, a bird!

Scatter

Did you know the moon does not produce its own light? What we see is sunlight striking the moon's rocky surface. The light scatters, or bounces off in many directions.

Transmission

Light is *transmitted* when it passes through a *transparent* object, such as a glass window. The flow of light is not interrupted.

Diffraction

When sunlight passes through a gap in the clouds, it bends through the gap and spreads out. This *diffraction* creates many beams of light shining toward the ground.

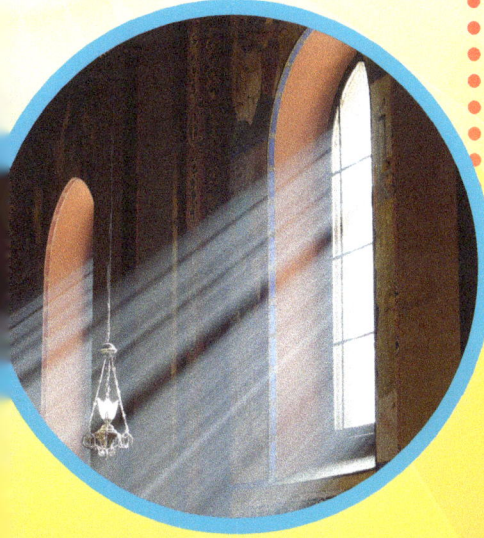

CURIOUS CONNECTIONS

ATMOSPHERIC SCIENCE

Earth is surrounded by a blanket of gases called the atmosphere. Blue light scatters as it moves through the atmosphere. This is why the sky looks blue.

2

SPECTRUM OF LIGHT

The electromagnetic spectrum is the range of all the different forms of light—both visible and invisible. And all this light moves like waves on a beach.

Radio waves

Microwaves

Infrared light

You're listening to pop songs on the radio as you microwave a bag of popcorn. Strange as it may seem, both your radio and your microwave use invisible light waves from the electromagnetic spectrum!

Invisible light waves include radio waves, microwaves, infrared light, ultraviolet (UV) light, X rays, and gamma rays. These forms of light are invisible to the human eye. Even though humans can't see these light waves, we have found ways to use them in various technologies.

X rays have an important role in health care.

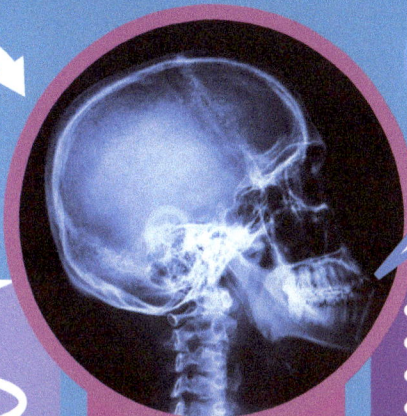

The X rays always find me.

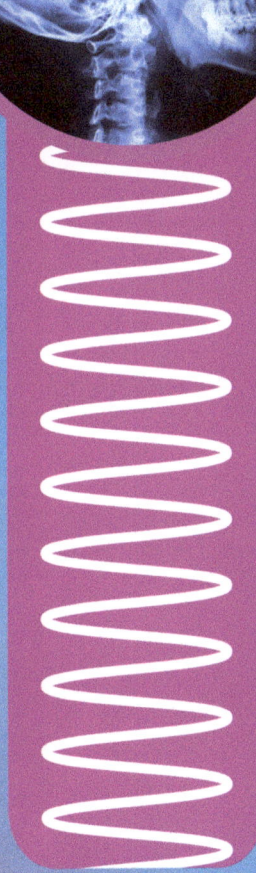

Visible light **Ultraviolet light** **X rays** **Gamma rays**

Visible light is the light people can see.

These light waves make up just a tiny portion of the electromagnetic spectrum.

Radio waves are used in communication.

What's in a **wave**?

Wavelength measures the distance between two similar points in a wave.
For example, wavelength is determined by measuring the distance between two adjacent high points of a wave.

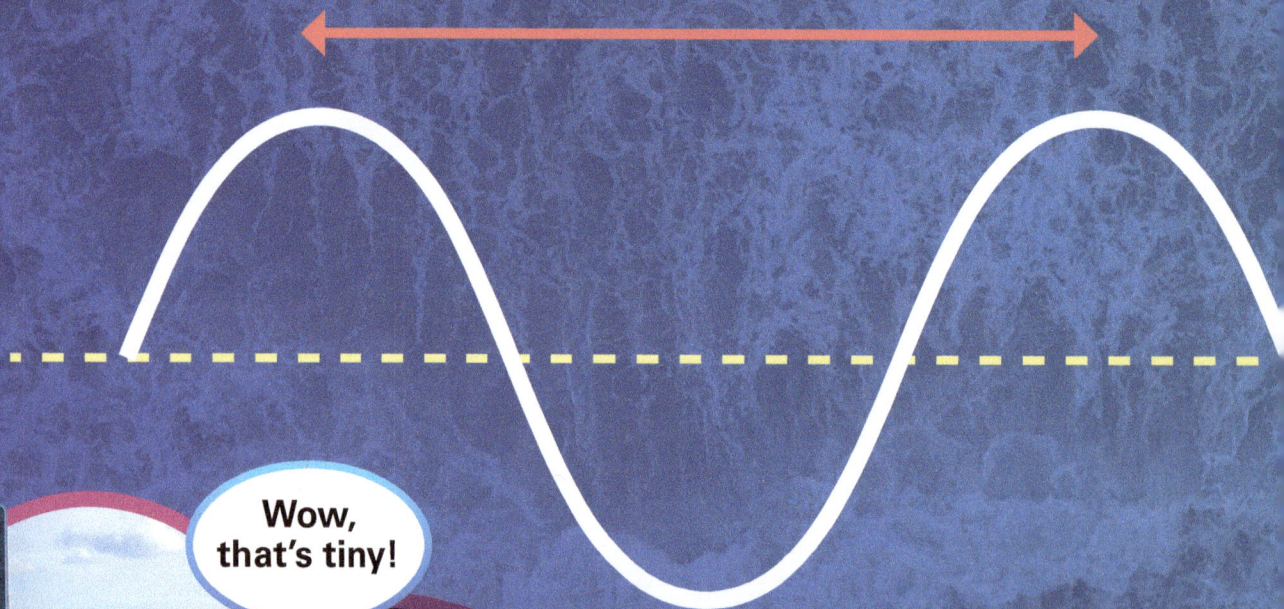

Wow, that's tiny!

Radio waves have the longest wavelength of all light in the electromagnetic spectrum—they can stretch for hundreds of miles! Meanwhile, the longest X rays have a wavelength that is about 10,000 times smaller than the width of a strand of hair!

Remember how light moves in waves?

Light waves have high points and low points, just like ocean waves.

Amplitude is the maximum distance the wave reaches from its resting point.

The resting point of this wave is shown by the dotted line. The amplitude is the measurement from the resting point to the highest or lowest point of the wave.

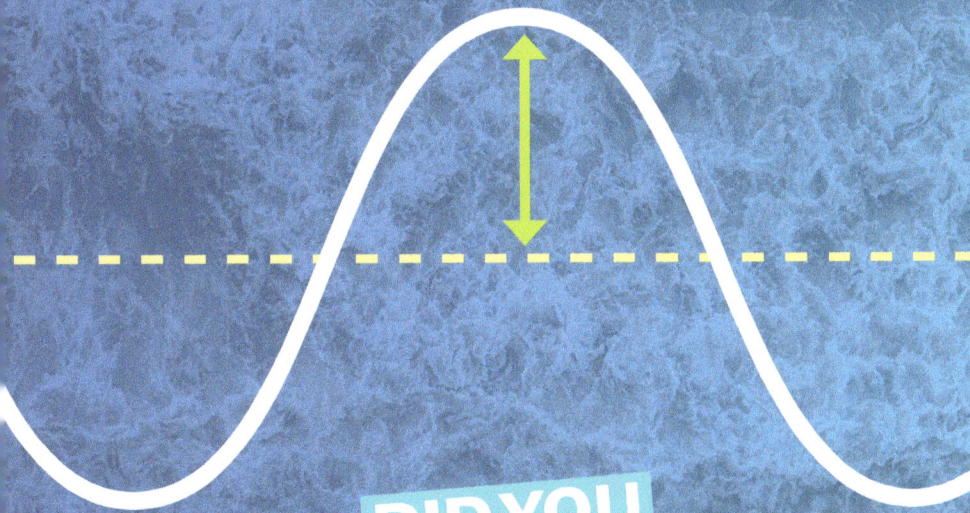

DID YOU KNOW?

Wavelength is measured in nanometers (nm). A nanometer is 10 million times smaller than a centimeter!

Amplitude is related to the brightness of visible light. Light waves from a bright light have a higher amplitude than light waves from a dim light.

Increasing **frequency**

Look at the forms of light in the illustration of the electromagnetic spectrum. Radio waves are at the far left end of the spectrum, and gamma rays are at the far right end.

Radio waves

Microwaves

Infrared light

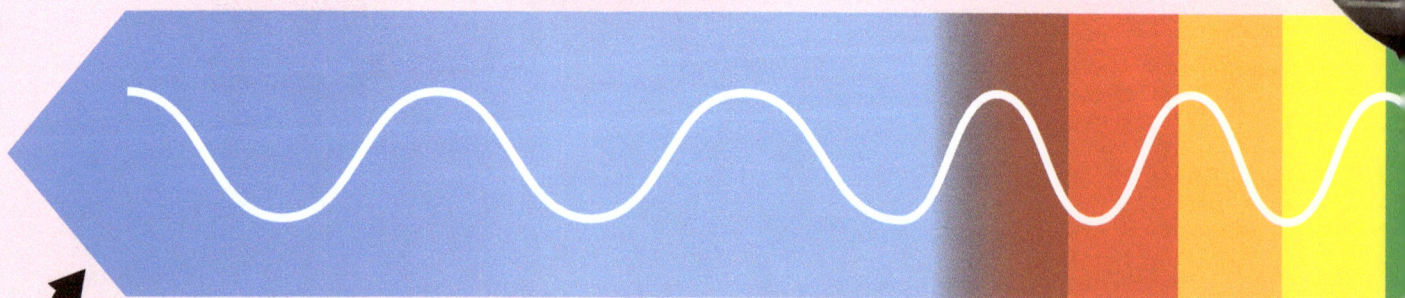

Radio waves have a long wavelength and a low frequency.

Wavelength determines the color of visible light. Red light has the longest wavelength. Violet light has the shortest wavelength.

Visible

In addition to wavelength and amplitude, a wave of light also has *frequency*. The more energy a light wave has, the higher its frequency. To understand what this means, let's compare different forms of light!

Imagine a radio wave and a gamma ray are walking together at the speed of light. The radio wave has a long wavelength, so it takes huge steps. The gamma ray takes tiny steps. To keep pace with the radio wave, the gamma ray needs more energy so it can take steps more frequently. In other words, the gamma ray has a higher frequency!

Ultraviolet light

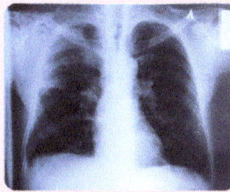

X rays

Gamma rays

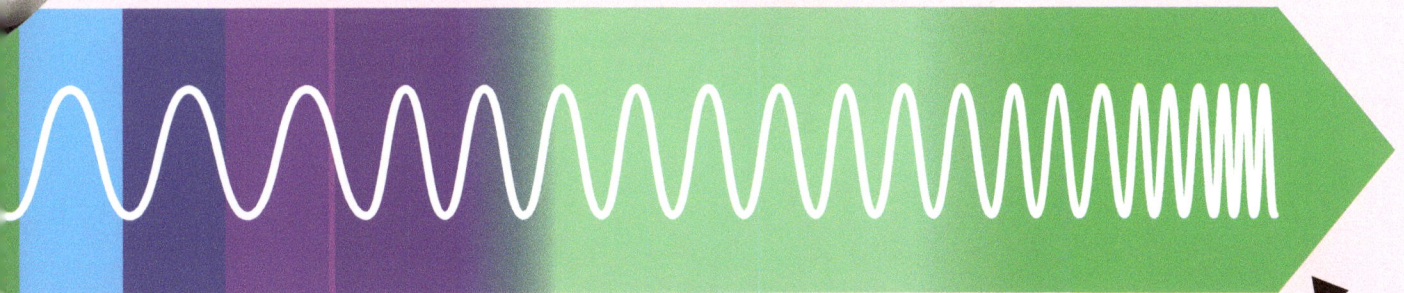

Gamma rays have a short wavelength and a high frequency.

spectrum

Getting **warmer**

I forgot my sunscreen!

Different light waves give off different amounts of radiation. UV light, X rays, and gamma rays are all high-energy light waves. They give off radiation that can be harmful to your body. For example, UV radiation from the sun is what causes sunburn. On the other end of the spectrum, such low-energy light waves as radio waves do not give off enough radiation to affect your health.

Nothing ruins a day outside like a terrible sunburn. Ouch! This happens because sunlight gives off energy called *radiation*.

Heat, like light, is a form of radiation. When light interacts with any material, the *electrons* in the material vibrate. This causes the material's *atoms* to move and collide with each other. We feel this motion as heat!

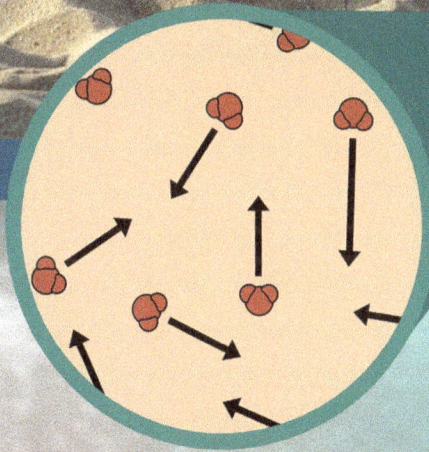

CAREER CORNER

Exposure to some types of radiation can damage *cells* in the human body and cause cancer. But radiation can also be used to kill cancer cells. Doctors called radiologists use machines to send radiation directly to a tumor.

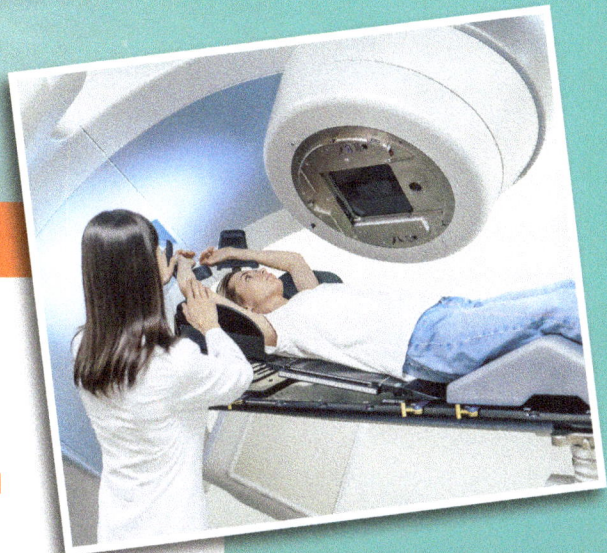

23

VISIBLE LIGHT

Visible light is the portion of the electromagnetic spectrum you are probably most familiar with. Neon signs, TV screens, holiday lights—these all give off light that can be seen with the naked eye.

Remember the name ROY G. BIV! It stands for the colors of the rainbow: red, orange, yellow, green, blue, indigo, and violet. These colors mix to form white light. The sun and light bulbs give off white light.

A *prism* is made of *transparent* material. When white light passes through the prism, the light bends, or *refracts*, into a rainbow that spills out from the other side of the prism.

Rainbows are created when sunlight interacts with water droplets in the air. The droplets are like tiny prisms that refract the light, separating it into colors we can see from the ground!

A rainbow always follows the ROY G. BIV pattern. This is because each color of the rainbow has a different wavelength (remember, red light has the longest wavelength and violet light has the shortest). Because of their different wavelengths, the colors of white light travel through the prism at slightly different speeds. This causes the light to bend at slightly different angles and separate into bands of color.

Mix and **match**

The primary colors of light are red, blue, and green. Secondary colors form when two of the primary colors interact. Red (R) mixed with blue (B) creates magenta (M). Blue and green (G) make cyan (C). Green and red create yellow (Y). And what if you shined all three primary colors of light at the same spot? They would merge to make white light!

R

Y

M

G

C

B

Painters can use red, blue, and yellow paints to create any color. Light is different! Red, blue, and green combine to make unique colors.

There is no unique wavelength that produces white light or black light. White light is the sum of all wavelengths of light, while black is the absence of light.

An object appears white when it *reflects* all light. It does not *absorb* any light.

An object appears black when it absorbs all light. No light reflects off the object.

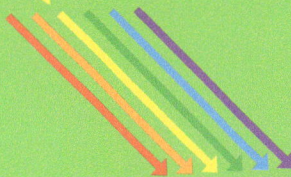

TECH TIME

The screen of a smartphone is made up of pixels. Each pixel can emit red, blue, and green light. The combination and intensity of these lights is responsible for creating the vivid colors displayed on the screen.

I spy . . .

When white light strikes an object, some wavelengths of the light are *absorbed*. Other wavelengths are reflected. The reflected wavelengths are the colors we see. Most objects reflect a combination of different wavelengths.

This strawberry reflects red light.

This traffic cone reflects orange light.

This leaf reflects green light.

Why is it difficult to see at night? Because there's less light! We see objects because they give off their own light or they *reflect* light. There's no sunlight at night. There's also less light reflecting off surfaces. If light doesn't reach our eyes, we can't see.

Vision begins with light entering the eyes.

Many parts of the eye work together to gather information about shape and color. Nerves in the eyes send this information to the brain. The brain processes this information and creates the images we see!

Cornea

Light enters the eye through the cornea. This is a clear protective layer that covers the eye.

Pupil

Light travels through the pupil. This is the opening to the interior part of the eye.

Lens

The lens is behind the pupil. It focuses light on the retina at the back of the eye.

Retina

The retina contains nerves that sense light and color. Rods are nerve *cells* that help us see when there is little light. Cones are nerve cells that detect color.

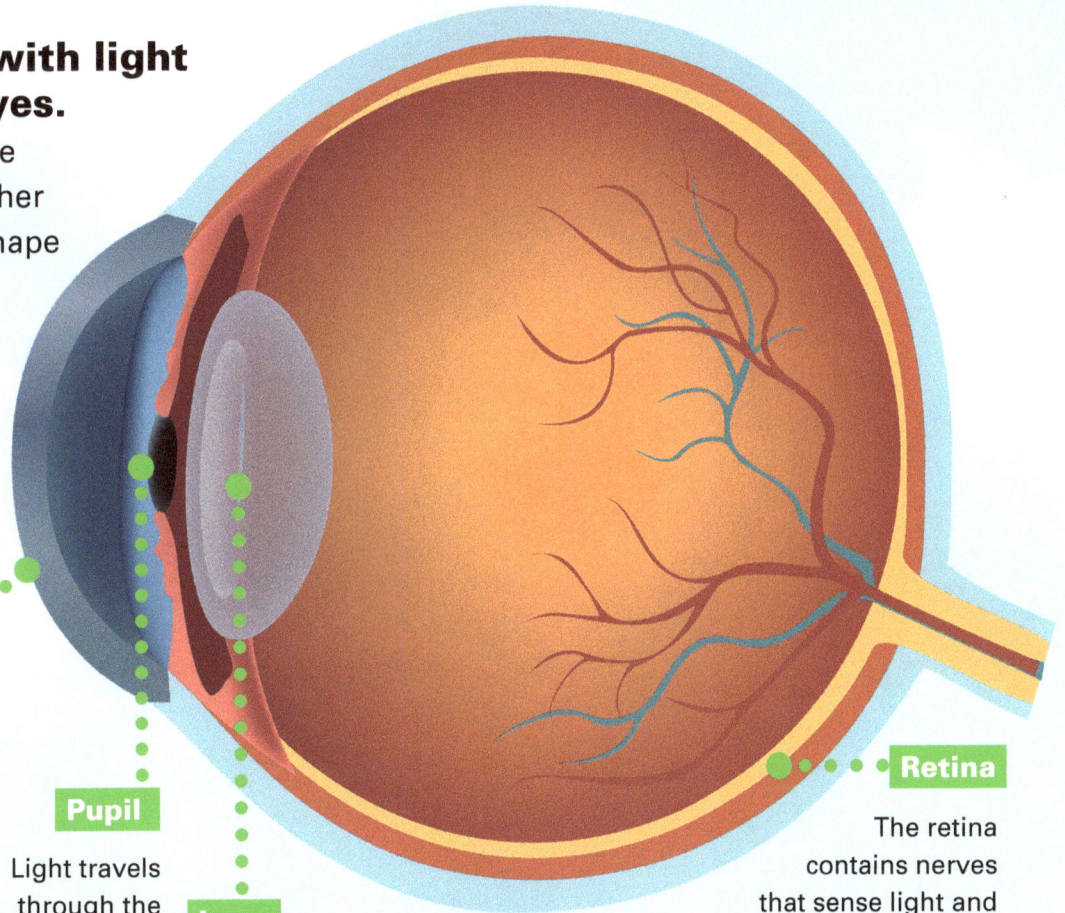

CURIOUS CONNECTIONS

BIOLOGY The eye has three types of cone cells. Each detects one of three colors: red, blue, or green. Do you have trouble seeing the number in this color blindness test? If so, your cones that detect red or green light may be missing or less sensitive.

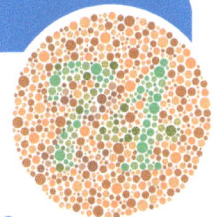

29

INVISIBLE LIGHT

Some types of invisible light have wavelengths that are longer than visible light. These include radio waves, microwaves, and infrared light. Other types of invisible light have shorter wavelengths than visible light. These include UV light, X rays, and gamma rays.

Sss-seen anything cool lately?

What do snakes and butterflies have in common?

Both can see parts of the electromagnetic spectrum that are invisible to humans!

You may not know it, but you use invisible light every day. Radio waves play a huge role in communication. You wouldn't be able to text your friend, use wireless headphones, or surf the web using Wi-Fi without radio waves!

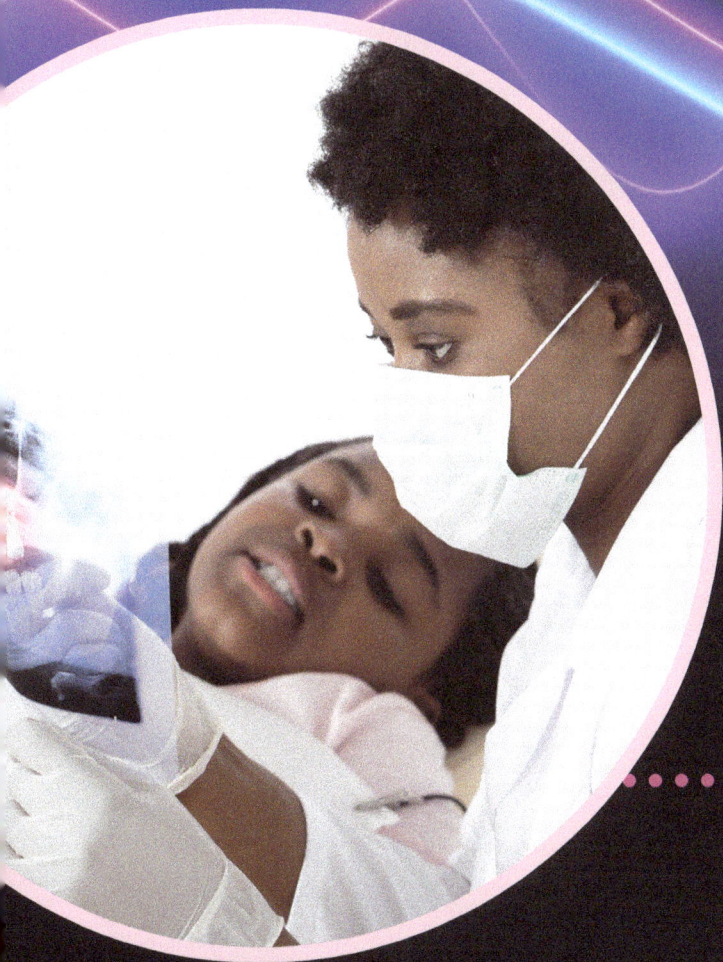

Health care wouldn't be the same without technology that uses invisible light. UV light can be used to clean surfaces and kill germs. And X rays allow doctors to take images of bones and tissues inside the body!

Over and **out**

Most radio waves pass through the atmosphere without being *absorbed*. Because of this, satellites in space use radio waves to send information to Earth.

CAREER CORNER

Scientists who study stars and other objects in outer space are called astronomers. Astronomers often use radio telescopes to locate faraway stars. Antennas on radio telescopes pick up radio waves that stars emit. Telescopes that rely on visible light don't work when it's cloudy. But radio telescopes work no matter the weather!

Radio waves have the longest wavelengths of all types of light.

Some radio wavelengths are as long as the height of the pyramids in Egypt. Others are much longer!

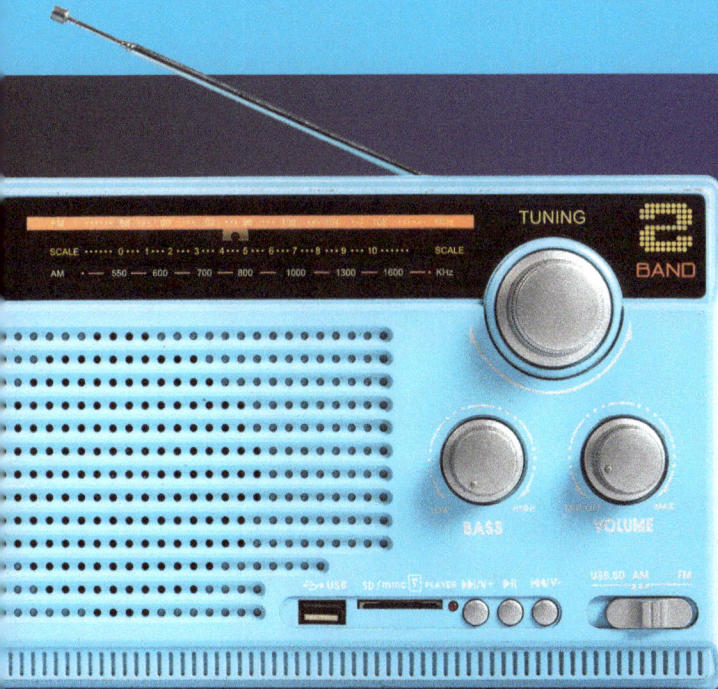

Radios use radio waves to function.

There are two main types of radio: FM and AM. FM radio stations and AM radio stations both tune to a specific *frequency* of radio waves.

FM radios

FM stands for frequency modulation. Information, such as music or a person's voice, can be *transmitted* using this type of signal. Adding this information changes the frequency of the radio wave. FM radio waves are typically used to transmit music because they can produce clearer sounds than AM radio waves.

AM radios

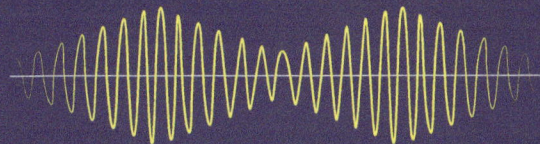

AM stands for amplitude modulation. AM radio waves tend to have lower frequencies than FM radio waves. When information is transmitted using an AM signal, the amplitude of the radio wave changes. AM radio waves can be transmitted over a greater distance than FM radio waves.

33

Dinner in **60 seconds**

Some wavelengths of microwaves can pierce through the atmosphere. These microwaves are not affected by weather conditions and can help *transmit* information between Earth and satellites in space.

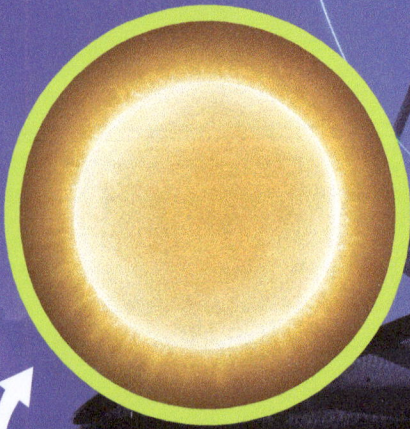

Stars are natural sources of microwaves.

TECH TIME

Radar technology relies on microwaves. Meteorologists use it to track weather patterns and storms. Radar is also used to detect the speed of vehicles and locate objects underground!

Need some quick heat for your cold soup?

Pop it in the microwave oven! This device is named after the light waves it uses to heat food. But microwaves come in handy beyond dinnertime!

Microwave ovens are designed to reheat food.

Food *absorbs* microwaves, which creates heat that cooks the food. Microwaves can pass through most materials, including glass, paper, and plastic. But metal *reflects* microwaves. This is why you shouldn't put metal objects in a microwave oven. The light waves will reflect off the object and bounce around. This can cause sparks and may start a fire!

1. A part of the microwave oven known as the magnetron produces microwaves.

2. Microwaves reflect off the metal interior of the device and hit the food.

3. Microwaves interact with water inside the food. The water begins to vibrate, producing heat.

4. The heat cooks the food from the outside to the inside.

POWER

TIMER

35

Feeling the **heat**

All objects give off some amount of infrared light.
But hot objects, such as flames, give off more infrared light than cool objects.

You can't hide from me!

Humans cannot see infrared light without special technology. But some animals can detect this type of light. This ability helps such fish as salmon hunt in murky waters!

Near and far: These words describe distance, but they are also types of infrared light! Far infrared light has longer wavelengths and can be felt by humans as heat. Near infrared light has shorter wavelengths and cannot be felt by humans.

Remote controls use near infrared light. When you press the power button on your TV's remote control, it gives off pulses of infrared light. A receiver in the TV then decodes the pulses as a command to turn the TV on or off.

Infrared light helps astronomers identify stars by their temperature. Red dwarfs are the coldest type of star. Blue stars are the hottest.

TECH TIME

Humans can detect infrared light with the help of a thermal imaging camera. This device detects very small differences in temperature. It displays these different temperatures as different colors on a screen, creating an image humans can see. Thermal imaging is used to track animals at night, detect electrical issues, and more.

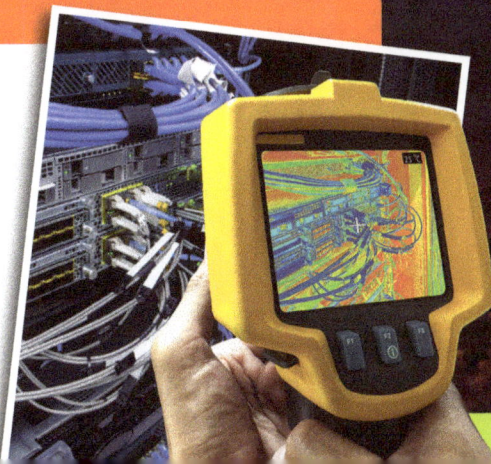

Fun in the **sun**

Earth's atmosphere *absorbs* much of the sun's harmful UV light. But a small portion of UV light reaches Earth's surface. UV light with the longest wavelengths is more likely to get through the atmosphere.

Long-term exposure to UV light can cause sunburn and some types of skin cancer. Sunscreen prevents most UV light from reaching the skin. But some amount of UV light gets past the sunscreen.

You might get a tan after spending a day outdoors, or you might burn. Either way, it's important to cover up with sunscreen to protect your skin!

UV light is invisible to humans, but some animals can see it! For example, UV light *reflects* off many types of flowers. Butterflies can see this UV light, which helps them find nectar to drink.

Did I miss a spot?

DID YOU KNOW?

The sun protection factor (SPF) determines the strength of sunscreen. The higher a sunscreen's SPF, the more UV light it blocks. When wearing sunscreen, a person can spend a longer time in the sun without burning.

A look **inside**

X rays are mostly used for medical imagery. Say you broke a bone in your leg. An X-ray technician would position your leg on top of a special plate. Then a machine would send X rays toward your leg. The *radiation* would pass through your leg and get *absorbed* by the plate.

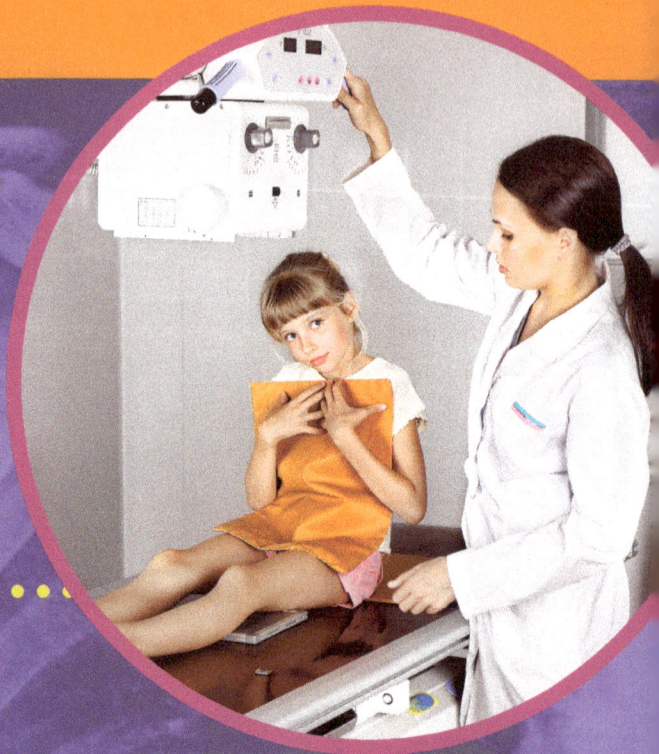

German scientist **Wilhelm Conrad Röntgen** discovered X rays in 1895. One of the first X-ray images he took was of his wife's hand! Röntgen won a Nobel Prize for Physics in 1901 for his discovery.

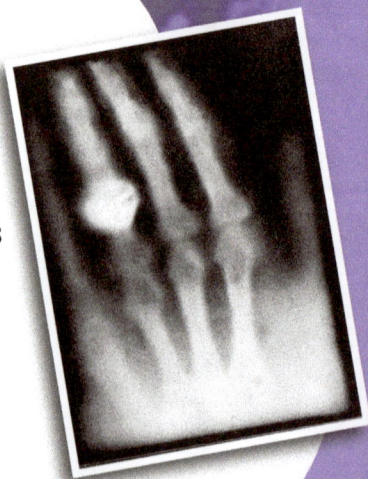

Have you ever gotten an X ray when you visited the dentist or doctor? X rays are like magic beams of light. They let us see what our eyes can't—including the bones inside our bodies!

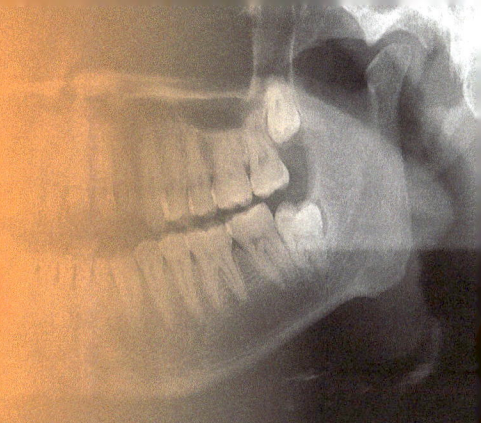

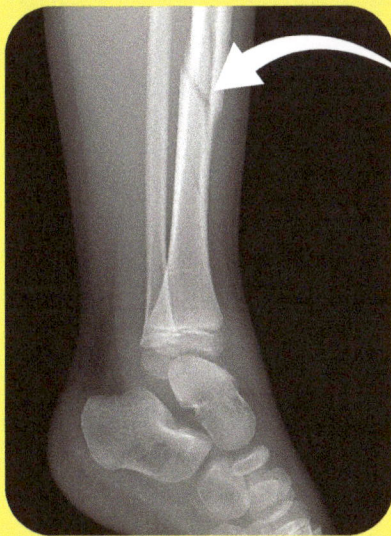

So how do X rays produce an image? X rays easily pass through skin, muscle, and fat before getting absorbed by the plate. These areas appear darker in an X-ray image. Meanwhile, bones are dense, making it difficult for X rays to pass through them. These areas appear lighter in an X-ray image. But if you had a broken bone, the X rays would pass through the crack in your bone. You would see the break as a dark line on your bone!

Radiologists use X rays to see a person's internal organs. This allows the doctors to locate tumors and diagnose diseases. X rays are also commonly used to kill cancer *cells*.

Massive **energy**

The hottest parts of the universe are natural sources of gamma rays. An exploding star—called a supernova—ejects massive amounts of gamma rays. Lightning also produces gamma rays.

Gamma rays, anyone?

Nuclear reactions produce gamma rays too. These reactions take place when the center of an *atom*, called the nucleus, is split. This releases a huge amount of heat and energy.

42

Only the hottest and most energetic objects in the universe produce gamma rays. The *radiation* from this type of light is extremely dangerous!

Nuclear power plants harness this energy to generate electricity. Equipment in nuclear power plants keeps nuclear reactions under control. Gamma rays produced by nuclear reactions are harmful to humans, so nuclear power plants undergo regular safety inspections.

CURIOUS CONNECTIONS

ASTRONOMY Gamma rays teach us about the universe. For example, a black hole is an area in space where gravity is so strong that even light cannot escape its pull. Black holes produce gamma rays. Satellites have detected gamma rays from a black hole that is 12.8 billion light-years away!

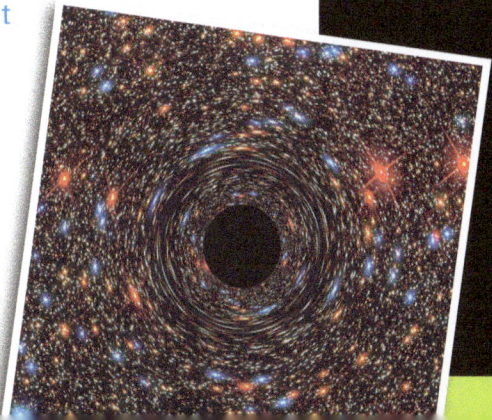

Refraction rainbow

You will need:

- A clear glass baking pan
- Water
- White, red, and blue paper
- Sunlight
- An outdoor surface, such as a table, and a friend to help hold the baking pan

Give it a try

1. Fill the baking pan about halfway with water.
2. Place the pan so part of it is extending over the edge of the outdoor surface. Make sure the pan does not fall over! Your friend can help hold the pan in place.
3. Place the sheet of white paper on the ground beneath the pan. Slowly angle the pan until a rainbow appears on the paper. Try to make the rainbow as large as possible. What colors do you observe in your rainbow?
4. Place your blue sheet of paper on top of the white sheet. How did this change the rainbow? Which colors do you see now?
5. Remove the blue sheet of paper. Then place the red sheet of paper on top of the white sheet. Observe how this changed the rainbow. What colors do you see now?

Weather conditions need to be just right for rainbows to form.

It can also be tricky to see the unique colors of the rainbow before it fades from the sky. But with some water and a clear pan, you can *refract* sunlight to make your own rainbow!

Try this next!

Draw a rainbow on a sheet of paper, arranging the colors from left to right. Then fill a clear glass with water and look at the rainbow through the glass. Notice the order of the colors when you look at them directly versus through the glass of water. Draw a sketch that shows how refraction is responsible for this effect.

QUESTION TIME!

How do you think the order of colors in a rainbow relates to the wavelength of each color? Which color has the longest wavelength? What about the shortest?

Index

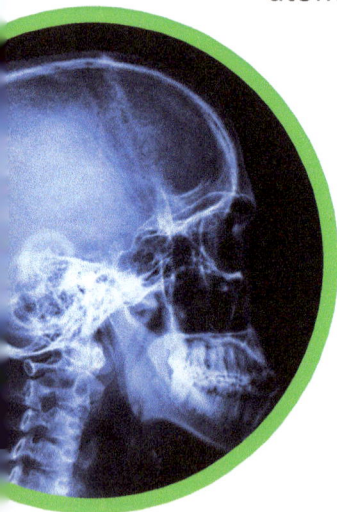

Glossary

absorb (ubb ZORB)—to soak up

atom (AT um)—the smallest unit of matter

cell (SEL)—a tiny unit of life that makes up all living things

diffraction (dih FRAK shun)—the bending and spreading out of light around an obstacle

electron (ee LEK trahn)—a negatively charged particle that moves around the core of an atom

frequency (FREE kwen see)—the number of waves that pass a certain point in a period of time

mass (MAS)—a measure of the amount of matter in an object

matter (MA ter)—the material that makes up physical objects

opaque (oh PAKE)—not see-through

photon (FOE tahn)—a tiny, massless particle that makes up light

prism (PRIH zem)—a piece of glass or other transparent material that is cut in a way that allows it to bend white light and cause a rainbow to form

radiation (ray dee AY shun)—energy that spreads out from a source in the form of waves

reflect (rih FLEKT)—to bounce or cause to bounce off an object

refract (rih FRAKT)—to bend or cause to bend

translucent (trans LOO sent)—partially see-through

transmit (trans MIT)—to transfer or pass from one place to another

transparent (trans PARE ent)—fully see-through

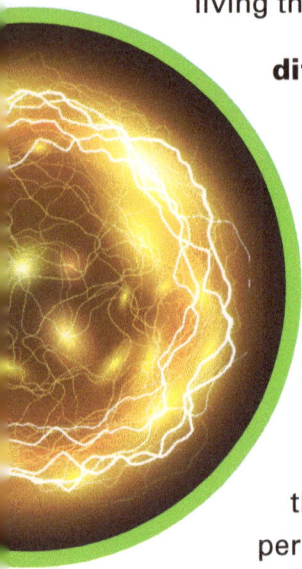

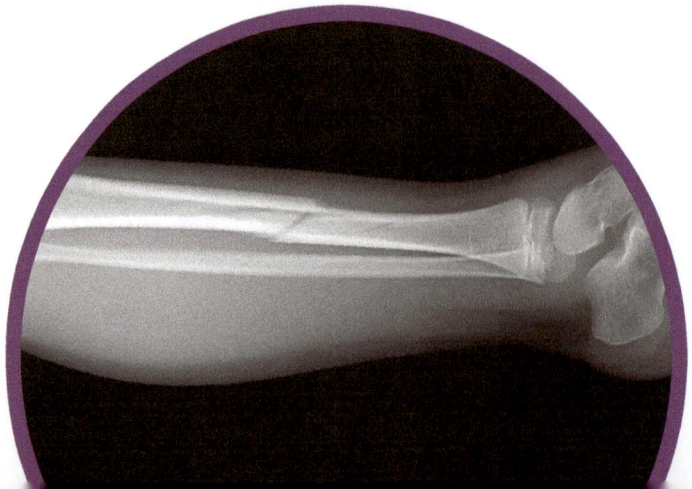

www.ingramcontent.com/pod-product-compliance
Lightning Source LLC
Chambersburg PA
CBHW040144200326

41519CB00032B/7595